Ecosystems Research Jo...

The Great Barrier Reef Research Journal

Natalie Hyde

CRABTREE
Publishing Company
www.crabtreebooks.com

Crabtree Publishing Company
www.crabtreebooks.com

Author: Natalie Hyde

Editors: Sonya Newland, Kathy Middleton

Design: Rocket Design (East Anglia) Ltd

Cover design: Margaret Amy Salter

Proofreader: Angela Kaelberer

**Production coordinator and
    prepress technician:** Margaret Amy Salter

Print coordinator: Margaret Amy Salter

Consultant:

Written and produced for Crabtree Publishing Company
by White-Thomson Publishing

Front Cover:

Title Page:

Library and Archives Canada Cataloguing in Publication

CIP available at the Library and Archives Canada

Library of Congress Cataloging-in-Publication Data

Names: Hyde, Natalie, 1963- author.
Title: Great Barrier Reef research journal / Natalie Hyde.
Description: New York, New York : Crabtree Publishing Company,
    2018. |
Series: Ecosystems research journal | Includes index.
Identifiers: LCCN 2017029304 (print) | LCCN 2017030907 (ebook) |
    ISBN 9781427119308 (Electronic HTML) |
    ISBN 9780778734703 (reinforced library binding : alkaline paper) |
    ISBN 9780778734956 (paperback : alkaline paper)
Subjects: LCSH: Great Barrier Reef (Qld.)--Environmental
    conditions--Research--Juvenile literature. | Biotic communities--
    Research--Australia--Great Barrier Reef (Qld.)--Juvenile literature.
    | Ecology--Research--Australia--Great Barrier Reef (Qld.)--
    Juvenile literature. | Coral reef ecology--Research--Australia-
    -Queensland--Juvenile literature. | Great Barrier Reef (Qld.)--
    Description and travel--Juvenile literature.
Classification: LCC GE160.A8 (ebook) |
    LCC GE160.A8 H93 2018 (print) | DDC 577.7/89476--dc23
LC record available at https://lccn.loc.gov/2017029304

Crabtree Publishing Company

www.crabtreebooks.com   1-800-387-7650   Printed in Canada/082017/EF20170629

Published in Canada
Crabtree Publishing
616 Welland Ave.
St. Catharines, Ontario
L2M 5V6

Published in the United States
Crabtree Publishing
PMB 59051
350 Fifth Avenue, 59th Floor
New York, New York 10118

Published in the United Kingdom
Crabtree Publishing
Maritime House
Basin Road North, Hove
BN41 1WR

Published in Australia
Crabtree Publishing
3 Charles Street
Coburg North
VIC, 3058

Contents

Mission to the Great Barrier Reef

Marine science has always been fascinating to me as a researcher. Imagine how excited I was to receive an email today from a conservation organization called Reefs Alive! I am heading to Australia to study one of the most amazing natural structures in the world. The Great Barrier Reef is the world's largest **coral reef** ecosystem. I have been asked to investigate how the reef is holding up against threats to its survival. Some of those threats are:

- **coral bleaching**
- outbreaks of crown-of-thorns sea stars
- damage to coral from fishing equipment and storms
- habitat loss from new developments
- increased ship traffic.

I plan to volunteer on some new projects that study and record changes to the reef. My trip will take me from the southern tip of the Great Barrier Reef all the way up to the Lizard Island Research Station.

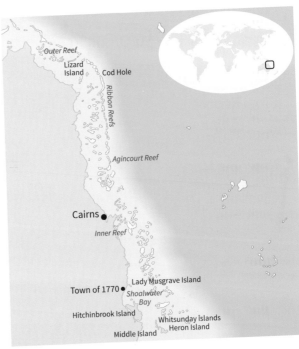

I was surprised to learn that the Great Barrier Reef is made up over 2,900 individual reef systems and over 900 islands. Reefs are ridges of living corals in the sea. The Great Barrier Reef stretches more than 1,400 miles (2,300 kilometers) up the east coast of Australia. Inner reefs are found close to the shore. They are about 115 feet (35 meters) deep. Outer reefs grow where the seabed begins to drop much lower. The reefs farther out from shore can be up to 6,500 feet (2,000 meters) deep! Most of the reefs are protected in an area called the Great Barrier Reef Marine Park (GBRMP). The diversity of life here is the reason it was named a **World Heritage Site** in 1981.

The Great Barrier Reef is one of the Seven Natural Wonders of the World.

Lady Musgrave Island

I got my first view of the Pacific Ocean as I drove down Captain Cook Drive into the town of 1770. The southern tip of the Great Barrier Reef lies beneath its brilliant blue waters. I took a **catamaran** to Lady Musgrave Island. It took 75 minutes to cross the choppy water, but it was worth it! Lady Musgrave is the only island in the Great Barrier Reef that has a **lagoon** deep enough for boats.

Aboriginal people and Torres Strait Islanders, such as the Ngaro and Gurang tribes, are the only people allowed to use the resources of the Great Barrier Reef. They hunt, fish, and trade goods along traditional routes.

I met with park rangers stationed on the island. They watch over green and loggerhead sea turtles during the breeding and nesting season. The rangers make sure visitors to the beach do not disturb the turtles. Hatchlings can lose their sense of direction and die on the beach if they are picked up or touched. To protect them, the island is closed to tourists from February to March, when the hatchlings make their way to the water. The rangers also count the nesting females each year to estimate the population.

Green turtle.

natstat STATUS REPORT ST456/part B

Name: Loggerhead sea turtle
(*Caretta caretta*)

Description:

Loggerhead sea turtles are the largest hard-shelled sea turtles. They return to the beaches where they hatched to nest. If sea turtles are disturbed as they leave the water to nest, they may turn around and not nest at all. Light and noise can also throw them off. Any human interference can drastically affect the number of hatchlings.

Attach photograph here →

Threats:
Loss of habitat due to development, pollution, and capture by fishing equipment.

Status:
Endangered internationally; threatened in the USA.

Field Journal: Day 2

Capricorn group and Heron Island

I headed out in the morning to the north end of the group of reefs and islands called Capricorn. I wanted to check on the seabirds that feed and nest in and around the Great Barrier Reef. Birds such as the wedge-tailed shearwater fly close to the water's surface and snatch fish, squid, and **crustaceans** from the water to eat. They dig **burrows** on the reef islands, where they return to nest year after year.

More than 75,000 black noddies make their home on Heron Island in the Capricorn group during the breeding season.

Commercial fishing is allowed in the area, but it must be controlled. Fishing equipment and anchors dropped from boats can damage the corals. Overfishing can be a problem, too. If the number of fish around the reefs drops, so does the number of birds that eat them. The Great Barrier Reef Marine Park Authority monitors the health of bird populations. It also tries to limit the number of visitors and the amount of **debris** from fishing activity.

natstat STATUS REPORT ST456/part B

Name: Capricorn silvereye (Zosterops lateralis chlorocephalus)

Threats:
Loss of habitat due to storms and pollution.

Description:
This small, greenish bird lives on coral cays in the Capricorn and Bunker Reef groups of islands. Silvereyes breed regularly on Heron Island and mate for life with the same partner. They make a cup-shaped nest in thick bushes and lay between one and three pale blue eggs. Silvereyes eat insects and small **invertebrates** as well as fruit. The sweet sandpaper fig on Heron Island is an important source of food.

Status:
Least concern.

Attach photograph here ➡

9

Field Journal: Day 3

I could see sea snakes, like this banded sea krait, in shallow water.

Olive Head Point, off Middle Island

After a night camping on Lady Musgrave Island, I took a water taxi to Middle Island. This is just off Great Keppel Island. It was low tide, and the water had drained farther back off the shore. The reef was exposed, so I was able to walk on the sandy paths between the sensitive corals. I made sure I did not touch the plants or animals. I could still see sea snakes swimming in the ankle-deep water around hard corals. I also spotted clams and sponges clinging to the reef.

Sightings

I noticed a spaghetti worm (loimia medusa). Found in shallow reef areas, it spreads out its tentacles to catch passing food.

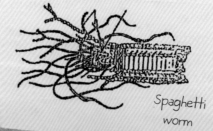

Spaghetti worm

There was also a lot of trash. Fishing nets, fishing lines, lumber, and other debris were choking the reef. There had been a bad storm a few nights ago. Strong waves and winds can push garbage dumped in the ocean to shore. Storms can even break the corals.
I saw a team of volunteers on a cleanup operation, and I offered to help. By the time the boat came to take me back to shore, the reef was in much better shape.

Trash is damaging the reef.

natstat STATUS REPORT ST456/part B

Name: Maxima clam
(*Tridacna maxima*)

Description:
These clams are the smallest relatives of the giant clam. They grow to 14–16 inches (35–40 centimeters) in size. The main color of their **mantle** is often blue, green, or purple. They bring water into and out of their valves, or shells, to get oxygen to breathe and **algae** to eat. They attach themselves loosely to rocks or reefs with long, tough fibers.

Attach photograph here ➡️

Threats:
Aquariums seek out these clams because of their beautiful colors and small size.

Status:
Least concern.

11

Field Journal: Day 4

The strange-looking dugong is also known as a sea cow.

Sailing to Shoalwater Bay

I wanted to visit a different part of the Great Barrier Reef Marine Park. Today I went to Shoalwater Bay. This area consists of large meadows of seagrass in shallow waters. Although there are no corals here, the animals that live in this area also use the reef habitat. Green sea turtles and other marine animals feed on the seagrass, but return to the reef beaches to nest. Fish and shrimp use the seagrass meadows as a nursery, or a place to raise their young. It is also home to the endangered dugong.

Shoalwater Bay is also the site of a military base. Armed forces from all over the world carry out military training in this sensitive part of the Marine Park. Leaders of the armed forces meet with the Great Barrier Reef Marine Park Authority twice a year to talk about the environment of Shoalwater Bay. Their activities affect the ecosystem, so the military operates a "no footprint" policy here. This means that all equipment, supplies, and trash are removed after training. Any ruts, tracks, or dirt mounds are restored to their original state, so there is "no footprint" of the activities left behind.

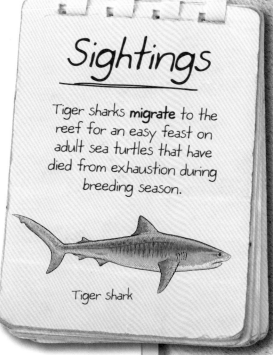

Sightings

Tiger sharks **migrate** to the reef for an easy feast on adult sea turtles that have died from exhaustion during breeding season.

Tiger shark

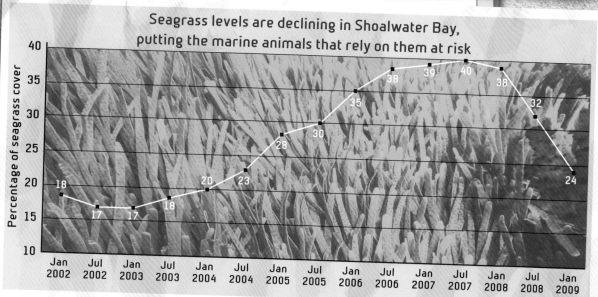

Seagrass levels are declining in Shoalwater Bay, putting the marine animals that rely on them at risk

Percentage of seagrass cover

Jan 2002	18
Jul 2002	17
Jan 2003	17
Jul 2003	18
Jan 2004	20
Jul 2004	23
Jan 2005	28
Jul 2005	30
Jan 2006	35
Jul 2006	38
Jan 2007	39
Jul 2007	40
Jan 2008	38
Jul 2008	32
Jan 2009	24

Anemone.

Snorkeling around the Whitsunday Islands

I could not have picked a more beautiful area to go **snorkeling**. The 74 Whitsunday Islands are in the heart of the Great Barrier Reef. Most of the islands are national parks that do not have people living on them. I have joined a group heading to Bait Reef. I had my first close-up look at the reef habitat when I slipped into the water. Colorful fish of all types and sizes live among purple, yellow, and red corals. I saw a turtle munching on an anemone and a reef shark circling, looking for lunch.

Sightings

I was lucky enough to see a flying fox. These are actually large fruit bats that roost in the trees on the islands.

Flying fox

After lunch, we moved to another section of the reef. It could not have been more different! The coral was ghostly white, and there were hardly any fish. This is a result of something called coral bleaching. Algae living in the coral give the coral its color. The coral cannot support both itself and the algae when the water is too warm. The coral forces the algae out. This causes the coral to turn white. **Climate change** is contributing to the warming of Earth and its oceans. This has led to an increase in coral bleaching. Coral can survive bleaching if it is not too bad, but often it leads to the death of the reef.

We measured the bleached coral to keep track of how widespread the problem is.

Mass coral bleaching events on the Great Barrier Reef

Year	Percentage of reef affected
1998	42%
2002	54%
2015	93%

Percentage of reef affected

Field Journal: Day 6

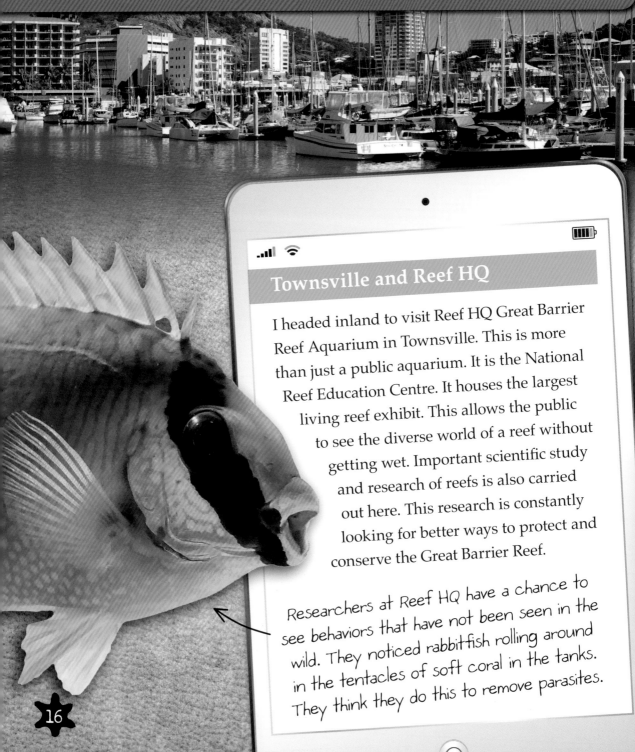

Townsville and Reef HQ

I headed inland to visit Reef HQ Great Barrier Reef Aquarium in Townsville. This is more than just a public aquarium. It is the National Reef Education Centre. It houses the largest living reef exhibit. This allows the public to see the diverse world of a reef without getting wet. Important scientific study and research of reefs is also carried out here. This research is constantly looking for better ways to protect and conserve the Great Barrier Reef.

Researchers at Reef HQ have a chance to see behaviors that have not been seen in the wild. They noticed rabbitfish rolling around in the tentacles of soft coral in the tanks. They think they do this to remove parasites.

Scientists are observing tanks of **venomous** fish such as stonefish and lionfish to find out how reefs deal with stress. They are also trying to establish the perfect conditions for coral to spawn, or lay eggs. Their goal is to have coral spawning as often as possible in nature. This will help the Great Barrier Reef recover from damage and rebuild the **colonies** of coral.

natstat STATUS REPORT ST456/part B

Name: Stonefish

Threats:
Loss of habitat, overfishing to use as aquarium pets and as food in some countries in Asia.

Description:
Stonefish live in coral reefs. They sit in or around rocks and plants. They blend in so well that they are hard to see. Stonefish are carnivorous, which means they eat meat, such as small fish and shrimp. They protect themselves with venomous spines. They are the most venomous fish in the world, and their sting can be fatal to humans.

Status:
Not enough data.

Attach photograph here ➤

Field Journal: Day 7

Mangroves are a linking ecosystem between land and reefs. Healthy mangroves help create a healthy reef.

Hinchinbrook Island mangroves and salt marshes

I drove north from Townsville to Lucinda, where I hired a boat to take me through Hinchinbrook Channel. This is a maze of waterways through mangroves and salt marshes between the coast and Hinchinbrook Island. Mangroves are flooded forests of trees. Marshes are areas of soft, wet land. These wetland areas are important for the health of the reefs. They help filter **sediment** out of the water so it does not get dumped in the reefs. The roots of the plants also help keep the shorelines from eroding.

Sightings

I got closer than I should have to a saltwater crocodile! It is breeding season, so he was probably warning me that I was in his territory.

Saltwater croc

A lot of crabs means a healthy mangrove and salt marsh.

Crabs are very important to mangroves and salt marshes. When they dig their burrows, they make holes in the sea bottom. This helps to flush out salt in the soil from the tides. Crabs also increase the oxygen in the water. Pockets of air get trapped in their burrows. These holes also provide habitats for other creatures. Crabs are a link between the land and reef ecosystems.

natstat STATUS REPORT ST456/part B

Name: Dawson yellow chat
(Epthianura crocea)

Threats:
Habitat loss due to dams, cattle grazing, and development.

Description:
The male bird is mainly yellow with a bright golden-yellow head. Females are a paler yellow. Chats raise their young in saltwater grassland. Females lay two to three eggs. Yellow chats eat mostly insects in and around the salt marshes.

Status:
Critically endangered.

Attach photograph here ➡

International cruise ships, yachts, catamarans, and cargo ships all sail through the reef to deliver goods and passengers.

Cairns

I drove north to check out the Cairns seaport and the marina. Cairns is a port, which is a city where ships can dock. Ports in the Great Barrier Reef area have been growing at a rate of about 2% per year. I took a boat tour to see how increased ship traffic is affecting the plants and animals of the reef. I could see evidence of a recent oil spill as we sailed out of the harbor. There were black **oil slicks** on some of the island beaches. I also saw a seabird called a tern hopelessly trying to clean the oil off its wings.

Sightings

The southern royal albatross does not breed on the Great Barrier Reef. It uses the area as a feeding ground on its way back to nest in New Zealand.

Southern royal albatross

Back in the port, I spoke to officials. They told me it would take about two weeks to clean up the islands, reefs, and water from this small spill. They also explained that new laws were being passed to protect the reef. Ships can no longer dump **dredge** material at sea. There will be no development of new ports in the Great Barrier Reef World Heritage Area. The existing ports and port traffic will also be better managed.

It can take weeks to clean up oil slicks on the beaches.

natstat STATUS REPORT ST456/part B

Name: Humpback whale
(*Megaptera novaeangliae*)

Description:
Humpback whales travel great distances through the world's oceans. They use their massive tail fin, called a fluke, to move through the water. They feed on tiny, shrimp-like krill, plankton, and small fish. They use the coral reefs as a quiet place to give birth. Once they regain their strength, they head out to open water with their calves.

Attach photograph here ➡

Threats:
Collisions with ships, entanglement with fishing gear, and noise pollution.

Status:
Once listed as endangered (1988), now considered least concern.

Field Journal: Day 9

Scuba diving the Outer Barrier Reef

The Outer Barrier Reef is the far edge of the reef that opens out to the ocean. This area supports different types of corals, fish, and **predators**. You cannot snorkel in water this deep, so I hired a boat to take me out to scuba dive in the Agincourt Reef. The walls of the reef here range from less than 3 feet (1 meter) deep to over 4,000 feet (1,200 meters). There were more hard corals than soft ones, since hard coral can survive the strong currents and tides. Those same currents help wash out the reefs and keep the water clean and clear.

The Agincourt Reef.

The Great Barrier Reef is the only living thing visible from space.

As I inspected the reef, I could see coral damaged by inexperienced or careless divers. Coral is fragile, and it can die if it is touched by divers or their equipment. But even these "dead" bits should not be touched. This is where new coral will try to grow. More areas of the Great Barrier Reef are being classified as "low density." This means that diving and fishing are being limited there to protect the reef.

I spotted a silvertip shark on the Outer Reef. These sharks can be dangerous to humans. I was glad it stayed far away.

Zones of the Great Barrier Reef Marine Park

General use 30%
Habitat protection 28%
Conservation park 4%
Marine National Park 33%

The four below make up less than 5% of the Marine Park

Preservation: a "no-go" zone, <1%
Scientific Research <1%
Buffer (no-go except for researchers) 3%
Commonwealth Island zones <1%

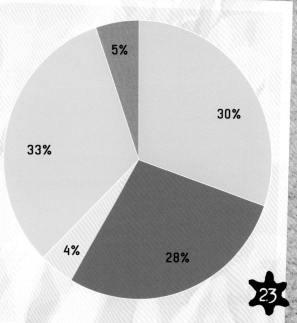

Field Journal: Day 10

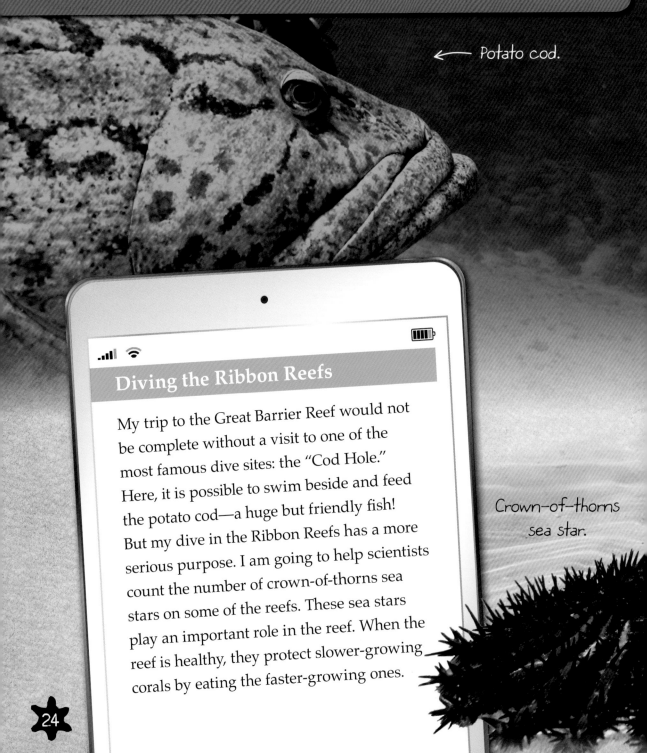

← Potato cod.

Crown-of-thorns sea star.

Diving the Ribbon Reefs

My trip to the Great Barrier Reef would not be complete without a visit to one of the most famous dive sites: the "Cod Hole." Here, it is possible to swim beside and feed the potato cod—a huge but friendly fish! But my dive in the Ribbon Reefs has a more serious purpose. I am going to help scientists count the number of crown-of-thorns sea stars on some of the reefs. These sea stars play an important role in the reef. When the reef is healthy, they protect slower-growing corals by eating the faster-growing ones.

However, a problem occurs when sea-star numbers increase or when the reef is stressed or unhealthy. At these times, the sea stars eat coral faster than it can grow. Flooding on the mainland can wash extra **nutrients** into the water. This extra food could cause a sea-star population explosion. More than 15 sea stars per 109,000 square feet (10,000 square meters) is considered an outbreak. Crown-of-thorns sea stars are the biggest threat to the survival of coral in the Great Barrier Reef.

If there had not been outbreaks of crown-of-thorns sea stars, there would have been an increase in average coral cover over the last two decades.

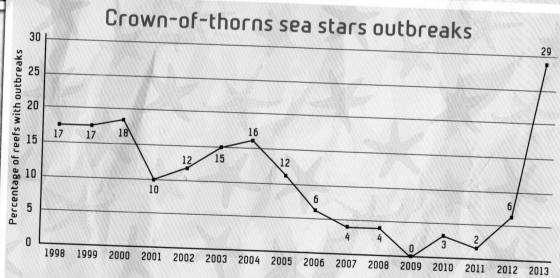

Crown-of-thorns sea stars outbreaks

Percentage of reefs with outbreaks

1998: 17
1999: 17
2000: 18
2001: 10
2002: 12
2003: 15
2004: 16
2005: 12
2006: 6
2007: 4
2008: 4
2009: 0
2010: 3
2011: 2
2012: 6
2013: 29

Field Journal: Day 11

Lizard Island Research Station

On my last day I flew to Lizard Island. It was at the research station here that scientists first learned that corals reproduce in one mass spawning per year. It only happens at night after a full moon. They also gained important information about coral trout. These fish spend most of their time in one small area except when they travel miles away during breeding season.

Over 100 projects researching the Great Barrier Reef are carried out from the station on Lizard Island each year.

Lizard Island was named by British explorer Captain James Cook for the number of monitor lizards that live there.

Scientists here are researching how climate change is affecting the reef. An increase in carbon dioxide in the air has changed the makeup of the water, making it more acidic. Researchers are doing experiments to learn what effect the higher acid content has on plants and animals. They have discovered that it makes it more difficult for corals to create their **exoskeletons**. It also makes it harder for fish to smell predators. Information that scientists discover here will help officials keep an eye on the health of the Great Barrier Reef.

Research here is helping put laws in place to protect this amazing ecosystem.

natstat STATUS REPORT ST456/part B

Name: Blue-spotted ribbontail ray
(Taeniura lymma)

Threats:
Habitat loss, fishing for use in aquariums.

Description:

This is a fairly small ray. It can be identified by the electric-blue spots on a yellowish background, and blue stripes on its tail. Rays feed at night near the shore and retreat to the reef during the day. Females give birth to up to seven young. Their tail spines can hurt humans, but rays would rather swim away than attack if they feel threatened.

Status:
Near-threatened.

Attach photograph here ⟶

27

Final Report

Report to: REEFS ALIVE!

OBSERVATIONS

My journey shows that the Great Barrier Reef faces many challenges. Some of those are natural, such as extreme weather, and some are due to human activity.

FUTURE CONCERNS

Storms such as cyclones are natural events that can do great damage to the reef. Outbreaks of crown-of-thorns sea stars is another. Rising water temperatures are putting the corals under stress. This causes coral bleaching, which can kill entire reefs. As climate change continues to increase water temperatures, more of the reef will die.

The communities of fish that rely on coral are affected by coral bleaching and the death of reefs.

There are now limits on putting up new buildings along shorelines.

The Great Barrier Reef Extreme Weather Program was created to help the reef recover after damaging weather events such as Cyclone Debbie in 2017.

Conservation Projects

Tourism and fishing in and around the reefs can cause damage to the delicate coral reefs, islands, and cays. Laws have been created to protect the structures and animals. In the Marine Park, different zones have been created to help limit or eliminate human activities in the reef (see page 23). Scientists recognize that mangroves, salt marshes, and seagrass meadows all play a part in keeping the reef water clean and reef creatures healthy. Studies being done at the Lizard Island Research Station are helping scientists and officials recognize threats to the reef and come up with new ways to protect this amazing ecosystem.

Your Turn

✳ Researchers and officials try to find a balance between letting the public enjoy the beauty and diversity of the Great Barrier Reef and protecting the reef and the animals that live there. Create a brochure for visitors to the reef. Explain the wonders they can enjoy, while reminding them of how they can protect the reef with their actions.

✳ Bar graphs, line graphs, and pie charts are all visual ways information can be shown. Take the information in one of the graphs in this book, and write a report explaining what you learned from the data.

✳ Imagine you are doing research at the Lizard Island Research Station. Write a journal entry describing your experiment and what you might learn from it.

Learning More

BOOKS

Coral Reefs by Kristin Rattini (National Geographic Children's Books, 2015)

The Mystery on the Great Barrier Reef by Carole Marsh (Gallopade International, 2006)

Where Is the Great Barrier Reef? by Nico Medina (Grosset & Dunlap, 2016)

WEBSITES

https://climatekids.nasa.gov/coral-bleaching/
Play a game and learn about coral bleaching and water pollution in the reef.

http://www.ducksters.com/science/ecosystems/coral_reef_biome.php
Explore coral reefs with Ducksters.

http://squidsquad.com.au/reef-care.html
Squid Squad has information, games, and ways people can look after the reef.

http://video.nationalgeographic.com/video/oceans-narrated-by-sylvia-earle/
oceans-barrier-reef
Take a virtual tour of the Great Barrier Reef with National Geographic.

Glossary & Index

aboriginal people who have lived in a land from the earliest times, or before the arrival of colonists

algae a simple form of plant that lives in water

burrows holes dug by small animals or birds as a home or nest

catamaran a boat with two hulls

climate change a change in global climate patterns due to increased greenhouse gases, such as carbon dioxide, in the Earth's atmosphere

colonies areas of identical coral

commercial things that are done to make money

coral bleaching when corals lose the algae living in them and turn white

coral reef a ridge of rock in the sea formed by the growth and deposit of coral

crustaceans invertebrate marine animals such as crabs, lobsters, and shrimp

debris scattered pieces of trash

dredge dirt, weeds, and mud pulled out of a river, lake, or ocean bed

exoskeletons the hard outside of animals such as corals, crayfish, and beetles

invertebrates animals that do not have a backbone

lagoon an area of salt water separated from the sea by a sandbank or coral reef

mantle the outer layer of the body of animals such as clams

marine relating to the sea or oceans

migrate to move from one habitat or region to another

nutrients substances that provide nourishment

oil slick a layer of oil on the surface of water

predators animals that live by killing and eating other animals

sediment tiny pieces of solid material carried and deposited by water

snorkeling swimming just below the surface of the water, using a breathing tube and mask

venomous poisonous

World Heritage site a place that is considered to be of particular natural or cultural interest and that is specially protected or preserved